REPORTS FROM

THE BOTANICAL INSTITUTE, UNIVERSITY OF AARHUS

NO. 14

A Brief Introduction to

The AAU-Flora of Ecuador Information System

by

P. Frost-Olsen & L.B. Holm-Nielsen

1986

This brief introduction to the AAU Flora of Ecuador Information System was distributed at the Utrecht Symposium on Tropical Botany on October the 2nd 1986 during an online presentation of the database.

Peter Frost-Olsen & Lauritz Holm-Nielsen

ISSN 0105-4236
ISBN 87-87600-18-8

```
Please make your selection by typing
the proper F-key of your choice:


        F1  Collections
        F2  Localities
        F3  Collectors
        F4  Taxonomy
        F5  Labels
        F6  Retrieval
        F7  Duplicates
        F8  Help
        F9  -
        F10 Next menu

        F11 BYE
```

```
    Welcome  to  the
   FLORA OF ECUADOR
   information system
```

```
   One moment, please

   The databases are now
  being made ready for you
```

F-key 4U   pfo

Contents

Introduction

During the last eight years computer technology has been used in the Ecuador Project at the Botanical Institute, University of Aarhus.

The first ideas concerning the possible use of computers arose, when we needed a faster way to manage large amounts of new information and plant collections from Ecuador in the late seventies. We realized that by using the electronic possibilities we would be able to handle large part of the trivial work such as producing labels, keeping records of localities etc., in a fast and efficient way. Our hope was to use less energy on the data management, leaving more time and energy for the botanical research. Later computerized data management appeared not only to be a useful tool helping in doing the trivial work, but also to facilitate an easier and faster access to the data.

At an early stage of the computer project we discussed the advantages and disadvantages of being dependent on hardware outside the institute. We found that the risk of technical isolation and the possibility of not being able to keep up with hardware and software developments were much larger, if we chose to develop the planned system on a minicomputer within the institute. We still believe this to be true. From the beginning we chose to develop the system on the main-frame computer at the University of Aarhus in order to have access to enough storage capacity.

Concerning software we found that it should be possible to develop a better suited and much more efficient software than was commercially available at that time. Accordingly we entered the risk of developing our own database system and Peter Frost-Olsen was employed as a fulltime programmer of the system.

What started as a simple filing of locality data and number data in order to produce labels, soon developed into a larger and more complex database system, the AAU Flora of Ecuador Information System.

As is generally known, the information technology has exploded during these eight years. It has been a challenge, but also hard work keeping up with this development. However, we have now a functioning system available, which is an effective tool in the management of the original collections and in the floristic work of the Ecuador Project at the Botanical Institute. As a result of the latest development of the system we have a fast on-line multi-user system with the needed high degree of security. A brief outline of the system will be given below and presented at the Symposium of Tropical Botany in Utrecht.

During the development of AAU Flora of Ecuador Information System we have passed through several periods of frustration. Some of these relate to common problems and are of a general character:

    ° In the Ecuador group at the Botanical Institute most of the botanists have not had sufficient understanding of the logic of EDP. What seems to be an easy task for the botanist, can at times only be solved through complicated programming. — But what the botanist would not dare to ask, is presented as a logical and necessary part of the system.

    ° We have often seen discrepancies between the expectations of the botanists and what is possible for the programmer within a short period of time.

4

° In the computer world the tendency is either nothing or everything. A hardware or software problem tends to close the system: Either thousands of labels or no labels at all.

° Coordinating larger projects with many people involved around a common database demands a higher degree of consequence, consistency and planned working processes than is generally applied by field biologists. Problems with coordinating the working processes will of course not arise in single person projects using a microcomputer.

Nevertheless a number of frustrations has had external reasons:

° The operating system of the computer and the network connection with the host was changed two times during 1981-1986, resulting in situations like: "It worked yesterday, today the system is dead".

° We have experienced curious transmission problems between the Botanical Institute and the computer center.

These two types of problems have delayed us more than one year. We have now been promised two to three years of hardware and software stability on the main-frame. This period will be used to further development. The system, however, has to be transferred to a new VAX computer at the computer center before 1990. We do not expect the transference to create serious problems.

Being attached to a main-frame computer has forced us to keep up with the development in computers   This would not have been the case, if we had based our computerization on mini-computers. Then a new computer generation would not automatically be acquired at the Institute, due to objections from the granting authorities. Only economically well-established major institutions or research programs can expect to change the equipment as the computer technology in minicomputers evolves.

In spite of the difficulties in applying this new tool in botany we are convinced that our efforts will prove to be fruitful. Today we have a system that is totally adjusted to the work of the botanist.

The following recommendations to newcomers in the computer club should be given:

° Don't forget that you are a botanist. Don't try to be a programmer. It would be futile to try to transform ourselves into second-rate computer scientists.

° Build up the needed expert knowledge. Try to understand the programmer.

° Build up an attachment to i.e. some non-profit organization or department working with computer sciences in order to keep up with the fast technical development.

° Use standard hardware and software.

° If you plan to build up large databases to be used by several persons, make sure that your hardware facilities have more than enough capacity. It will be needed sooner than you realize, both to meet the future demands for storing data and to secure the efficiency of the system.

6

° Use the possibilities of the modern equipment: store your information as meaningful data, i.e. use full botanical names. No codes are needed anymore.

° Insist on good output quality. The labels shall last for centuries.

° Have the network possibilities in mind. Think of future communication between databases

The AAU Flora of Ecuador Information System.

The system is developed on the UNI•C (Aarhus) main-frame computer. It is divided in three levels. The user needs a valid username, a password and an open account to reach the first level. This level is the normal sign-on procedure found on all multi-user computers. If the user signs on with the proper values he will have access to the computer operating system.

The operating system immediatly presents to the user a menu describing the type of terminal that can be used. If the user has access to one of these terminal types he can enter level two. He will be put on line to the AAU Flora of Ecuador Information System. The system presents the main menu. (Fig. 1). From this level he can select any parts of the system. This is level tree.

Before the user is accepted on this level he must enter the AAU Flora of Ecuador Information System specific password. If the right password is entered the screen presents the program specific menu as shown on Fig. 9. The password control concerns the user and his actual legal access. Depending on his password he may update, correct information, draw maps, make searches, define output from the database, or any combinations of these.

The user has to define under which relation he wants to use the information present in the database which is now available to him. He needs to define his job and choose the program by using the function-keys as indicated on the screen.

The third level contains all published and unpublished informations of the Ecuador project. The system is extremely flexible and the user is able to manipulate the content and all other features of the database. By using one of the keys F1, F2, F3 etc. he controls that part of the system he is allowed to use.

Fig. 1

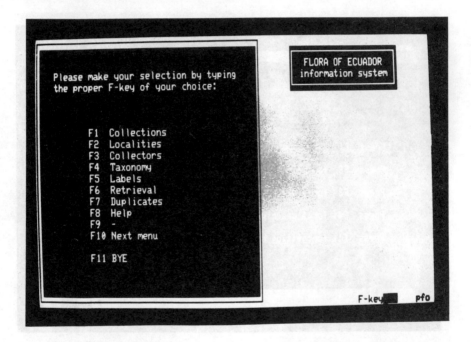

Please make your selection by typing
the proper F-key of your choice:

    F1 Collections
    F2 Localities
    F3 Collectors
    F4 Taxonomy
    F5 Labels
    F6 Retrieval
    F7 Duplicates
    F8 Help
    F9 -
    F10 Next menu

    F11 BYE

FLORA OF ECUADOR
information system

F-key   pfo

Main system menu. The AAU Flora of Ecuador Information System presents itself with this screen-form. All
programs or tasks a user may want to execute, shall be
selected from this menu. When a program or task is
finished, the system transfers control to this menu.

Fig. 2

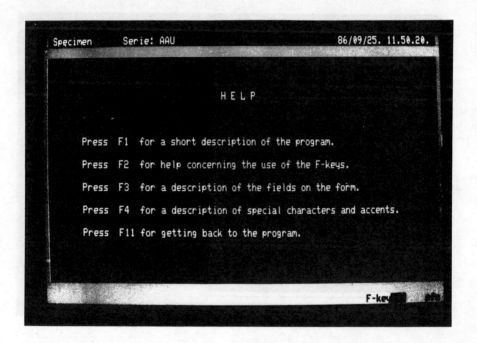

```
Specimen      Serie: AAU                          86/09/25. 11.50.20.


                            H E L P


        Press  F1  for a short description of the program.

        Press  F2  for help concerning the use of the F-keys.

        Press  F3  for a description of the fields on the form.

        Press  F4  for a description of special characters and accents.

        Press  F11 for getting back to the program.



                                                        F-ke
```

There are several kinds of on-line help available from the programs. When help is requested by a user, this screen-form is displayed.

Fig. 3

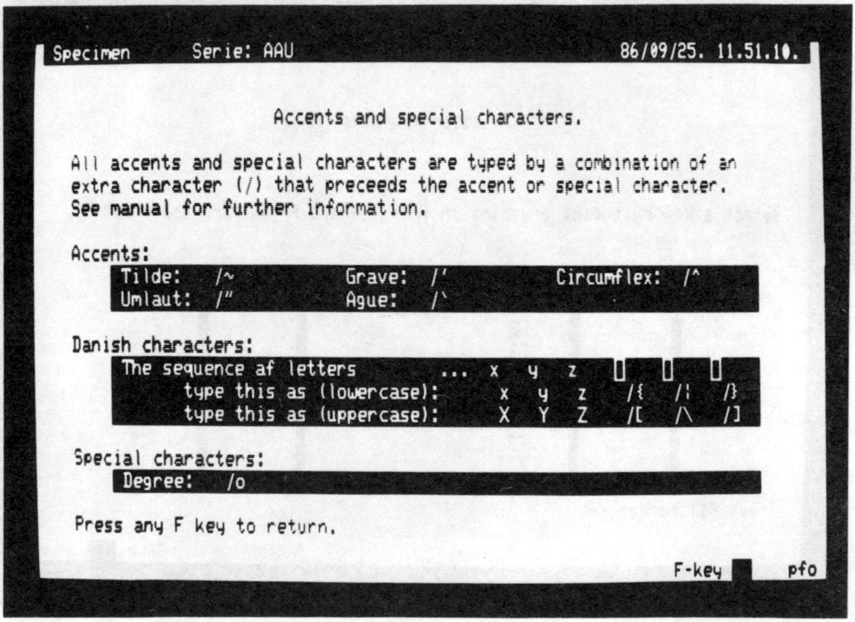

Accents and special characters.

All accents and special characters are typed by a combination of an
extra character (/) that preceeds the accent or special character.
See manual for further information.

Accents:
| Tilde: | /~ | Grave: | /' | Circumflex: | /^ |
| Umlaut: | /" | Ague: | /` | | |

Danish characters:
The sequence af letters	...	x	y	z	▯	▯	▯
type this as (lowercase):		x	y	z	/{	/¦	/}
type this as (uppercase):		X	Y	Z	/[	/\	/]

Special characters:
| Degree: | /o |

Press any F key to return.

F-key ▮   pfo

If the function key is pressed when the screen-form in
fig. 2 is displayed, this form appears. The screen-form
shows how to type special characters.

Fig. 4

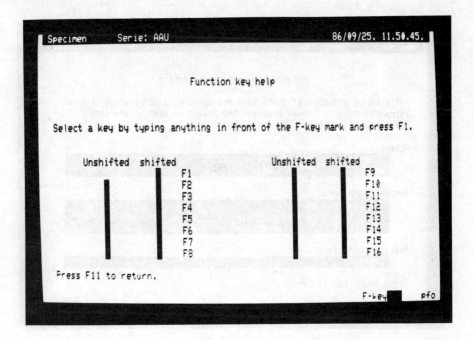

A user of the system shall never type a command. All information is entered by the user or displayed by the system on screen-forms. The terminal keyboard contains several special keys called function-keys. Each of these keys perform a special action, when pressed. A description of what action a function-key performs can be obtained by this form.

Fig. 5

```
 Specimen       Serie: AAU                        86/09/25.



                   Who are you ?  ████████
```

Several types of security control are used by the system.
When a program starts, this screen-form is always
displayed. The user can only continue with the program,
if he types a password in the dark field. Each user of the
system is provided with a personal password, describing
which program he or she may use and what he or she
may do in the program.

Fig. 6

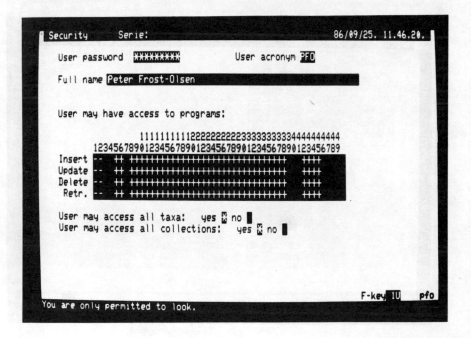

The password assignment is done by this screen-form.
Each user is identified internally in the system by the
password, externally by a user acronym.

Fig. 7

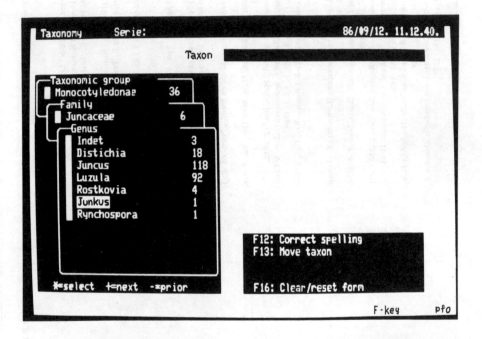

The database supports a full taxonomical subsystem. This part of the database is accessible by this screen-form. It is possible to view information concerning each taxon in the database as well as perform several kinds of updates and corrections of the taxonomy. The data on the form shows that there is one collection belonging to the genus *Junkus*. It is certainly a typing error for *Juncus*. By following instructions displayed in the lower right part of the screen-form when the key F12 is pressed, *Junkus* can be changed to *Juncus*.

Fig. 8

Balanophoraceae : *Corynaea*	Es	Ma	Gu	EO	Ca	Im	Pi	Co	LR	Bo	Tu	Ch	Cn	Az	Lo	Na	Pa	MS	ZC
Corynaea Hooker f.																			
crassa Hooker f., Trans. Linn. Soc. London **22**: 31, 54, *t. 13*. 1856.	.	.	.	.	.	.	+	.	.	.	+	.	.	+	.	+	.	.	.
≡ *Itoasia crassa* (Hooker f.) Kuntze, Rev. Gen. **2**: 590. 1891.																			
= *Corynae purdiei* Hooker f., Trans. Linn. Soc. London **22**: 31, 55. 1856.																			
≡ *Itoasia purdiei* (Hooker f.) Kuntze, Rev. Gen. **2**: 590. 1891.																			
= *Corynae sphaerica* Hooker f., Trans. Linn. Soc. London **22**: 31, 55, *t. 14*. 1856.																			
≡ *Itoasia sphaerica* (Hooker f.) Kuntze, Rev. Gen. **2**: 590. 1891.																			
Helosis L. C. Richard																			
cayennensis (Swartz) Sprengel, Syst. Veg. **3**: 765. 1826.	.	.	+	.	.	.	.	.	+	.	.	.	.	+	.	.	.	+	.
cayennensis (Swartz) Sprengel var. **cayennensis**	.	.	+	.	.	.	.	.	.	.	.	.	.	+	.	.	.	+	.
≡ *Cynomorium cayennense* Swartz, Nov. Gen. Sp. Pl. 12. 1788. [basionym]																			
≡ *Caldasia cayennensis* (Swartz) Mutis ex Steud., Nom. ed. 2. **1**: 255. 1840.																			
= *Helosis guyanensis* L. C. Richard, Mém. Mus. Hist. Nat. **8**: 416, 432, *t. 20*. 1822.																			
= *Helosis brasiliensis* Schott & Endl., Melet. Bot. 12. 1832.																			
≡ *Helosis guyanensis* L. C. Richard f. *brasiliensis* (Schott & Endl.) Eichl. in Mart., Fl. Bras. **4.2**: 23, *t. 5 f. 1*. 1869.																			
≡ *Caldasia brasiliensis* (Schott & Endl.) Kuntze, Rev. Gen. **2**: 590. 1891.																			
cayennensis (Swartz) Sprengel var. **mexicana** (Liebmann) B. Hansen, Bot. Tidsskr. **72**: 188. 1978.	.	.	.	.	.	.	.	.	+	.	.	.	.	.	.	.	.	.	.
≡ *Helosis mexicana* Liebmann, Forhandl. Skand. Naturf. 4. Möde 1844: 181. 1847. [basionym]																			
≡ *Caldasia mexicana* (Liebmann) Kuntze, Rev. Gen. **2**: 590. 1891.																			
= *Helosis guyanensis* L. C. Richard var. *andicola* Hooker f., Trans. Linn. Soc. London **22**: 57. 1856.																			
≡ *Helosis guyanensis* L. C. Richard f. *andicola* (Hooker f.) Eichl. in Mart., Fl. Bras. **4.2**: 23. 1869.																			
≡ *Helosis guyanensis* L. C. Richard var. *andicola* (Hooker f.) Eichl. in DC., Prod. **17**: 136. 1873.																			

Fig. 8. The taxonomy described in fig. 7 is a subset of a more elaborated taxonomic database under development. This database is used to store nomenclatural and distributional data and will be used for a general-purpose checklist producing information system. The figure shows the possibilities from the pilot version of this system. (Data from B. Hansen, Flora of Ecuador 19, 1985.)

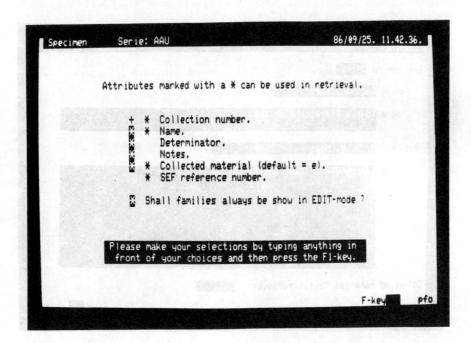

Fig. 9. When a password is accepted by the system, the user will be presented a program or task specific menu. The menu can offer the user several choices or options. A selection from the menu is made by filling out the form and then pressing a dedicated function key. In the example shown here the user has selected all choices except the SEF reference number by typing a * before all the available choices. Fig. 10 shows what happened after the function key was pressed.

Fig. 10

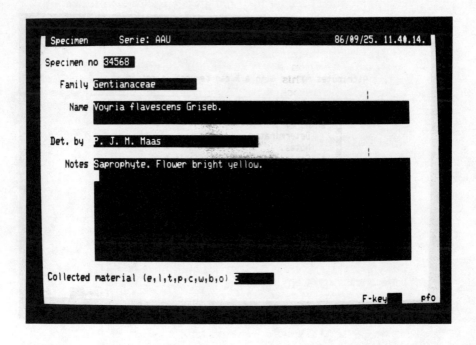

A new screen-form is displayed with several new fields. The fields selected in the former screen (fig. 9) are now displayed. In the present form (fig. 10) it is possible for the user to enter information or to display information from the database. The information shown in the example was obtained by typing the specimen number in the proper field and pressing function key 2 (F2). The key informs the program that the form shall be filled with the information belonging to this specimen number (in database terminology: the key). It does more than that: the function key change the database to edit- or update-mode. This means, that all the displayed information are locked for any other user until the user who entered the update-mode, leaves it again. This can be done either by pressing the shifted F2-key, which ends the edit mode without updating the database, or the F1-key. If F1 is pressed, the database will be updated with the changes made to the information on the form. If the update was successful, the form will be cleared and the database will be unlocked. All data stored or updated by such a form will be checked for correct syntax (a specimen number cannot begin with a character) and, if it is possible, whether the value is valid (is the family known by the taxonomy?). All errors are presented to the user in the lowermost line of the form.

This form is the most used form in the system: all new collections and all updating of determinations of the collections in the database are entered by this form.

Screen-forms like this one can also be used for searching in the database. For example, if someone needs to know, whether a liquid preserved collection of Passiflora is available, he can get the answer by typing Passiflora in the "Name" field and a L (the only code used in the system!) in the "Collected material" field, and then press the F5-key. If any collections are available they can be displayed one at a time by pressing the F4-key.

Fig. 11

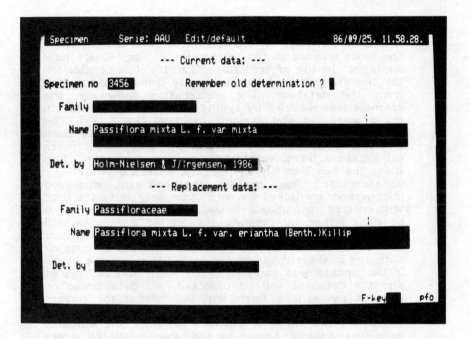

```
Specimen      Serie: AAU   Edit/default           86/09/25. 11.58.28.
                  --- Current data: ---
Specimen no  3456          Remember old determination ?
   Family
     Name Passiflora mixta L. f. var mixta

 Det. by Holm-Nielsen & J/irgensen, 1986
                  --- Replacement data: ---
   Family Passifloraceae
     Name Passiflora mixta L. f. var. eriantha (Benth.)Killip

 Det. by
                                                   F-key        pfo
```

When new determinations shall be entered in the database, it often occurs that several specimen numbers must be updated with the same name. Using the method described in fig. 10, implies that the determination must be entered for each specimen number separately. It is a rather slow method and it garanties typing errors when a determination is typed 10 or 20 times. If this situation occurs, a faster and simpler method exists: just fill out the form with the determination and press F8. This key informs the program to perform a syntax-check and validity check of the determination and enter the default mode. If the determination is accepted, the screen-form changes (fig. 12). The new form is used for entering the specimen numbers that shall have the new determination. When all these numbers have been entered the updating is initiated by pressing the F2-key.

Fig. 12

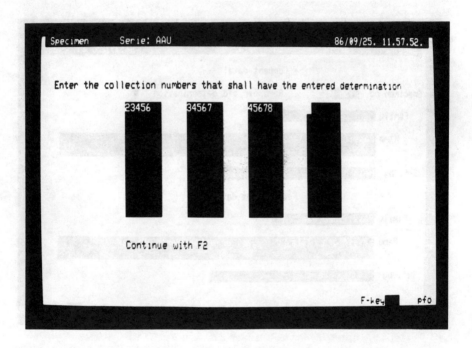

See fig. 11.

Fig. 13

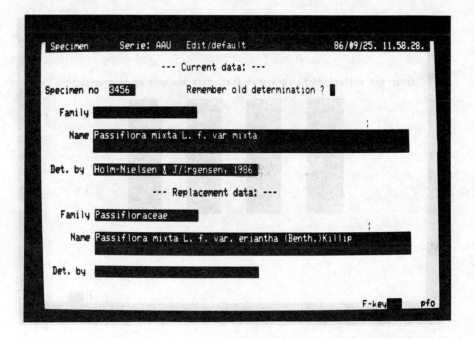

When the F2-key has been pressed the screen-form
changes to the present form (fig. 13). The lower half of
the screen displays the new determination that shall
replace the current determination, shown in the upper
half of the screen. The user must accept the change
before an update takes place. When all specimens numbers entered on the form shown in fig. 12 have been
updated the screen-form changes to the form that was
active when the F8-key was pressed.

Fig. 14

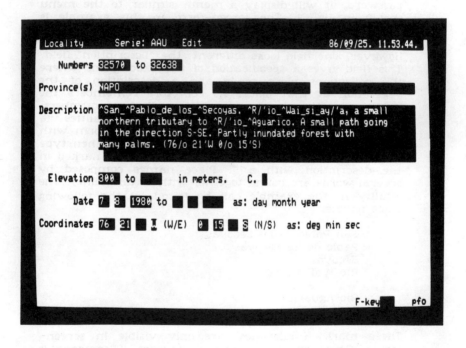

```
Locality    Serie: AAU   Edit                86/09/25. 11.53.44.

     Numbers 32570 to 32638

Province(s) NAPO

Description ^San_^Pablo_de_los_^Secoyas. ^R/'io_^Wai_si_ay/'a, a small
            northern tributary to ^R/'io_^Aguarico. A small path going
            in the direction S-SE. Partly inundated forest with
            many palms. (76/o 21'W 0/o 15'S)


  Elevation 300 to      in meters.   C.

       Date 7 3 1980 to           as: day month year

Coordinates 76 21    I (W/E)  0 15   S (N/S)  as: deg min sec




                                                F-key       pfo
```

Fig. 14

Fig. 9 to 13 deals with the specific information concerning the single specimen. A specimen is collected on a locality, often with several other specimens. Therefore the information concerning the localities is separately entered for each specimen. The locality program is selected in the main menu. Like all other programs this program also makes a request for a password. If it accepts the password, it will display a menu, similar to the menu shown in fig. 9. The screen-form in this example is selected by this menu. Most of the functions described in fig. 10 are available here too.

However, one field looks different: the "Description" field. This field gives a specification of the exact place where the collections were made and a description of the vegetation at that place. As collectors remembers the localities by the places and not the collections numbers, we have also wanted to have access to the localities via cited place-names. Instead of expanding the form with fields for entering the place names (you must then type the place names twice), the place names are marked in the description with a ^. Place names composed by several words are bound together with _ (underline). The locality in the example can be found by the following place names:

    San Pablo de los Secoyas
    Pablo de los Secoyas
    Secoyas
    Río Wai si ayá
    Wai si ayá
    Río Aguarico
    Aguarico.

These marks (underlines) are only visible in screen-forms. They are never printed (except if someone is wanting to see the marks).
This example also shows how accents or special characters are typed. Unfortunately it is not possible on the terminals we are using to type an accented e. Therefore some notation is necessary for entering such accented characters (fig. 3): you must type R/'io in the form in order to get Río printed on a label.

Fig. 15

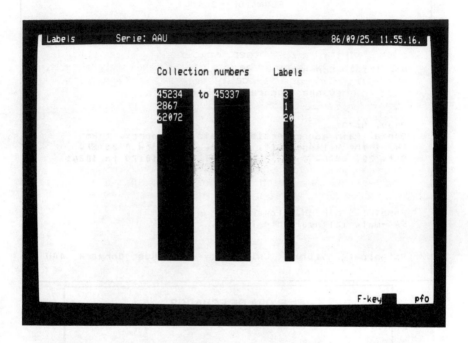

One very important output-product from the AAU Flora
of Ecuador Information System is the labels. All
collections made since the Third Expedition to Ecuador in
1976 have been labeled with labels produced by one of
the versions of the system. Fig. 16 shows the difference
between one of the first labels produced and a label of
the quality of today.

It is very easy to make a label: just enter the numbers
in the screen-form shown in the figure 15. The example
shows that 3 labels should be printed of each collection
made in the interval from 45234 to 45337, 1 label of 2867
and 20 labels of 62072. The labels will be produced and
printed immediately when leaving the program. The
labels are composed automatically from the information
in the database.

The security is also active with this program: if a user
isn't permitted to use the program, he or she cannot
produce any labels.

Fig. 16

```
                    FLORA OF ECUADOR


              Collected by B. Øllgaard & H. Balslev


No. 10180   Cannaceae

            Canna leucocarpa Bouché


Prov. NAPO:
Baeza. Path and riverside vegetation approx. 1 km
SW of the village. Alt. 2000 m.  (77 53'W 0 28'S).
Oct. 20, 1976.        Loc. incl. nos. 10179 to 10241




Herb, 1.5 m high. Flowers orange-red.
FAA-material available in AAU.


Botanical Institute  University of Aarhus  Denmark  AAU
```

```
                  FLORA OF ECUADOR
              Collected by L. P. Kvist & E. Asanza



No. 40810                              Bixaceae
     Bixa platycarpa

Prov. ESMERALDAS:
Río Cayapa, Zapallo Grande. In front of the village a
trail was followed into the forest. Disturbed forest.
(78°55'W 0°48'N) Alt. 100 m. 1-2 Aug. 1982.


Shrub, 4 m tall. In primary forest.
Nom. Vern.: Laj muchi (Cayapa);
Use: For colouring the food. The colour of the seeds
of this Bixa variety is yellow rather than red.

────────────────────────────────────────────────────────
Botanical Institute, University of Aarhus, Denmark (AAU). Project directed by L.B. Holm-Nielsen and B.Øllgaard
In collaboration with P. Universidad Catolica (QCA) and Museo Ecuatoriano de Ciencias Naturales (QNA), Quito.
```

Labels produced by the system.

Fig. 17

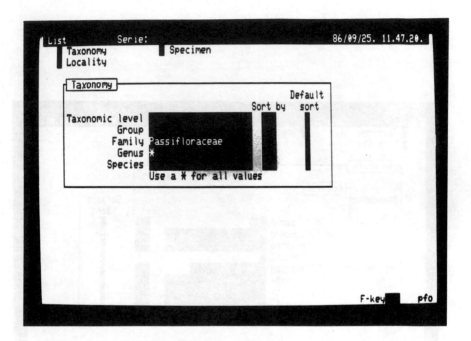

In fig. 10 it is told how the screen-forms can be used as a tool for searching for specific information. Results from these searches are only accessibly on the screen-forms that produced the result, the results are normally not printable. If printable output is wanted a general-purpose retrieval program must be used. This figure shows how this program functions. The user can select one or more "view"s of the database system. In this figure a view of the taxonomy is selected. The system displays a form with room for the user to enter the values or combination of values he wants to be used in his query. In the example the family Passifloraceae with all genera is wanted.

Fig. 18

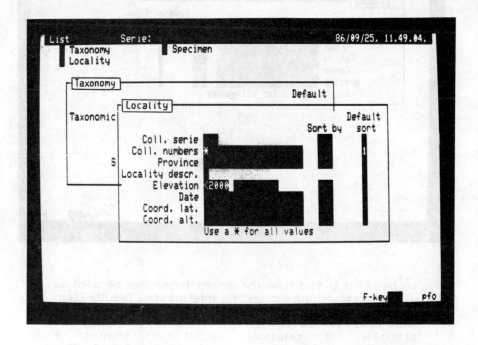

If the user wants the data obtained in the example in fig. 17 in combination with other information, he can select a further view. In this example he has selected the locality view. He now restricts the query to the taxa (genera in Passifloraceae) collected below 2000 m and the query shall search in all the localities (collections) in the database.

Fig. 19

```
Page 1      Program: List      Database ECUADOR        Output ==> PFO   86/09/17.
-------------------------------------------------------------------------------

Prov. NAPO:
Llanganati. Cushion paramo just below summit of Pan
de Azucar, dominated by Werneria humilis, Lycopodium
crassum and columnar Lachemilla. Soil temperature
6.5°. (78°18'W 1°9'S). Alt. c. 4100 m. 15 june 1982.

Collectors: B. Øllgaard, L. Holm-Nielsen, B. Boysen Larsen,
            L. P. Kvist, A. R. Jensen & S. Wium-Andersen


AAU  38540    E      Poaceae
                      Calamagrostis
                      Bunch grass.
AAU  38541    E      Poaceae
                      Festuca dolicophylla J. & C. Presl
                      Det. S. Lægaard
                      Bunch grass.
AAU  38542    E      Asteraceae
                      Loricaria complanata (Sch. Bip.)Wedd.
                      Det. H. Robinson
AAU  38543    E      Hypericaceae
                      Hypericum
                      Low shrub.
AAU  38544    E      Scrophulariaceae
                      Bartsia laticrenata Benth.
                      Det. N. H. Holmgren
AAU  38545    E      Poaceae
                      Agrostis tolucensis H.B.K.
                      Det. S. Lægaard
AAU  38546    E      Rosaceae
                      Lachemilla
AAU  38547    E      Rosaceae
                      Lachemilla nivalis (H.B.K.)Rothm.
AAU  38548    E      Ranunculaceae
                      Ranunculus guzmanii H.B.K.
AAU  38549    E      Gentianaceae
                      Halenia
AAU  38550    E      Gentianaceae
                      Gentiana sedifolia H.B.K.
```

Example of printable queries produced by the program described in fig. 18. This example shows all collections made on one locality. The determination, the fieldnotes and available material are printed for each collection.

Fig. 20

```
Page 1     Program: List        Database ECUADOR      Output ==> PFO   86/09/17.
------------------------------------------------------------------------------

   Triplaris dugandii Brandbyge

    AAU 32234
       Prov. MORONA-SANTIAGO:
       Huassaga. Primary forest with many palms. (77°12'W
       2°30'S) Alt. 300 m. 24 june 1980.

    AAU 32380
       Prov. MORONA-SANTIAGO:
       Pumpuentza. S-SW of the village. Disturbed primary
       forest. (77°20'W 2°25'S) Alt. 250 m. 29 june 1980.

    AAU 33304
       Prov. NAPO:
       San Pablo de los Secoyas. Path going in the direction
       NW. Very wet and partly inundated forest. (76°21'W
       0°15'S) Alt. 300 m. 7 august 1981.

    AAU 34607
       Prov. PASTAZA:
       Río Bobonaza. Rain forest on river bank and secondary
       forest around houses between Huagracachi and Cachitama,
       below Montalvo. (76°43'W 2°20'S) Alt. c. 300 m.
       18 july 1980.

    AAU 36208
       Prov. NAPO:
       Río Wai si ayá. Primary forest on non-flooded ground.
       (76°21'W 0°15'S) Alt. 300 m. 28 august 1981.

    AAU 58680
       Prov. NAPO:
       A⁻nangu, NW corner of the "Parque Nacional Yasuní".
       Undisturbed forest in the surroundings of the SEF terra
       firme line and KTH hectar plot. (76°22'W 0°33'S)
       Alt. 260-360 m. 1-30 april 1985.
```

Example of printable queries produced by the program
described in fig. 18. This example shows all localities
where a certain species have been collected.

Fig. 21

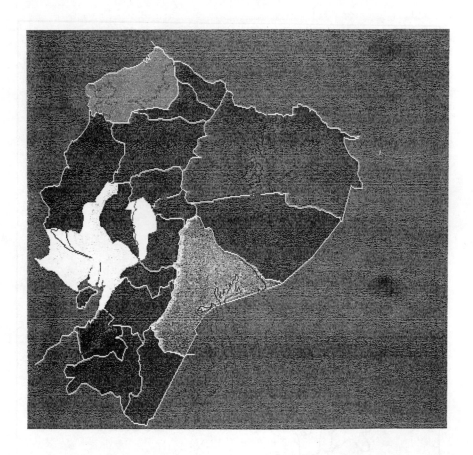

This figure shows a copy of a colour map produced by the system. The provinces are here seen in various shades of grey, in the original several colours are used for the provinces.

Fig. 22

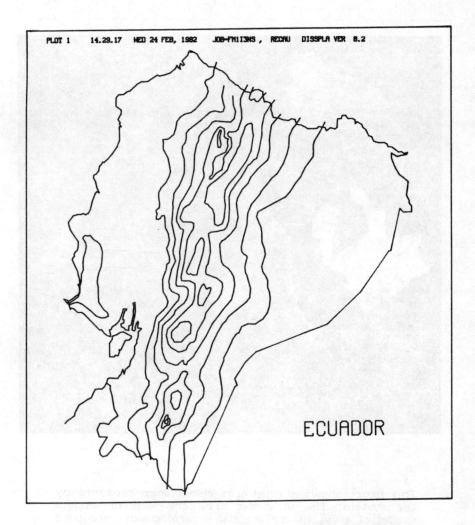

ECUADOR

Many different topological data have been digitized as
well as plantgeographical maps. All these data can be
displayed on the graphic terminals, but is also available
for normal plotting equipment. The figure shows
isotherms between 11°C and 23°C.

Fig. 23

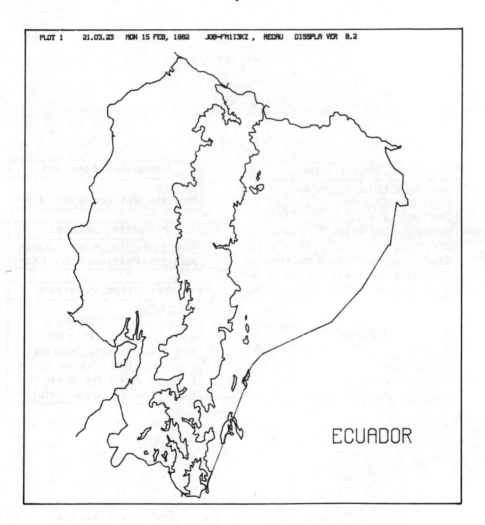

ECUADOR

Example showing the altitudinal curve of 1200 m plotted
on an Ecuador map.

Fig. 24

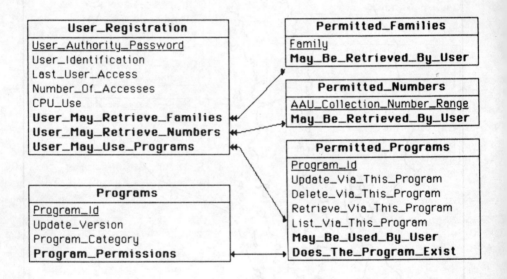

Simplified view on the structures (records) dealing with the security and control of the users access to the database. The arrows show the associations between the different structures.

34

Fig. 25

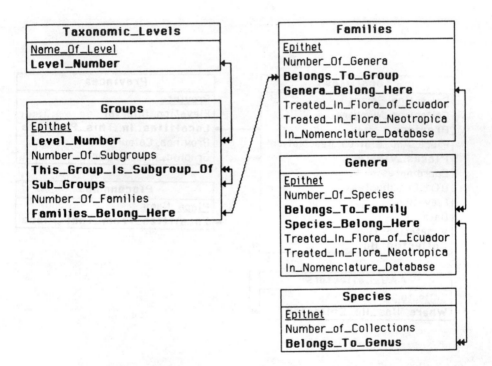

Simplified view on the structures dealing with the taxonomy.

Fig. 26

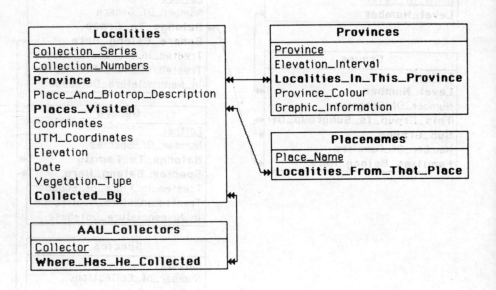

Simplified view of the structures dealing with the localities.

Fig. 27

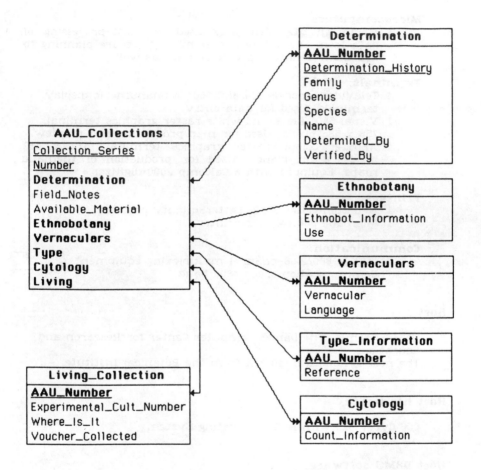

Simplified view of the structures dealing with the specimens.

Equipment at the Botanical Institute.

Microcomputers:
3 Macintosh 512. Primarily used for post-processing of data from the database system, but we are planning to use them as intelligent 'terminals' as well.

Terminals:
4 TeleVideo, models 924 and 950. Alphanumeric display terminals. Used for data entry.

1 Visual 550. Black and white raster graphics terminal, 758 x 585 pixels. Used for map production and preview.

1 AED512. Colour raster graphics terminal, 512 x 512 pixels, 8 bit planes. Used for production of thematic maps. Equipped with a CalComp 2000 digitizer.

Printers:
1 NEC Spinwriter 5510. Letter-quality printer.

1 Microline µ84. Matrix printer.

Communication:
2 MICOM 800/2. 8-channel multiplexing equipment. Communication speed 4800 baud.

Host.

UNI•C Aarhus, The Danish Computer Center for Research and Education.
The host is housed c. 10 km from the Botanical Institute.

Host hardware.

CDC Cyber 825 with NOS 2 Operating System.

Host DBMS software.

IMF 2.0 database management system.

AAU Flora of Ecuador software.

All parts of the AAU Flora of Ecuador system are developed at the Botanical Institute.
Programming language: PASCAL.

Acknowledgements.

The authors are indepted to the Botanical Institute, the Faculty of Science at the University of Aarhus and the Danish Natural Science Research Council for providing the necessary facilities. We are grateful to the students and staff of the AAU Ecuador group for discussions and for a lot of patience over the years. The staff of the computer center UNI●C (formerly RECAU) are thanked for fruitful interest and help in the project during the years and for quick access to new versions of software. The great help of Anni Slot and Kirsten Tind in preparing this Report is highly appreciated.

Selected Literature.

Abbott, L.A., F.A. Bisby & D.V. Rogers. 1985. - Taxonomic analysis in Biology; Computers, Models, and Databases. - New York.

Allkin, R. & F.A. Bisby. 1984. - Databases in Systematics. - London.

Brenan, J.P.M., R. Ross & J.T. Williams. 1975. - Computers in Botanical Collections. - London.

Date, C.F. 1981. - An introduction to Database Systems. - Mento Park.

Gómez-Pompa, A. et al. 1985. - Flora of Veracruz Project: An update on database management of collections and related information. - Taxon 34: 645-53.

Mackinder, D. & H. Synge. 1986. - Database News. - Threatened Plants Newsletter 15: 18-24.

Martin, J. 1981. - An end-user's guide to database. - New Jersey.